Student Activ

for use with

by

R. Thomas Wright
Professor Emeritus, Industry and Technology
Ball State University
Muncie, Indiana

and

Ryan A. Brown
Associate Instructor
Indiana University
Indianapolis, Indiana

Publisher
The Goodheart-Willcox Company, Inc.
Tinley Park, Illinois
www.g-w.com

Introduction

This Student Activity Manual goes hand-in-hand with the *Technology: Design and Applications* textbook.

Each Student Activity Manual section contains several activities. Some of them correspond to activities in your textbook, while others are independent of the text. Your teacher will give you more directions for completing these activities. Worksheets for the activities in your textbook are also included in this manual.

Each chapter also contains study questions corresponding to the material in your textbook. Your responses to these questions allow your teacher to know where you might need additional help. You should study the chapter material thoroughly before attempting to answer the study questions.

Safety should always be top priority in any activity. If you are not sure whether or not you are using materials and tools properly, check with your teacher. Always follow safety rules to avoid injury to yourself and others.

R. Thomas Wright

Ryan A. Brown

Scope of Technology

Name ______________________ Date ______________________

Period ______________________ Score ______________________

Activity 1-1
Using Technology to Make a Product

Making Tiles without Technology

Describe how you will make a tile from a lump of clay.

____

____

____

____

Product Analysis

Analyze the tiles members of your class made by completing this section.

Productivity

A. Number of tiles the class produced _____

B. Number of workers (class members) _____

C. Productivity of the work force (output divided by workers—A/B) _____

Quality

Write a statement describing each factor.

A. Uniformity:

- Thickness ______________________
- Shape ______________________

B. Surface smoothness ______________________

C. Grooves:

- Position of the tile ______________________
- Size (width and depth) ______________________

Making Tiles with Technology

In the left column of this form, list the steps you will use to make the product, as your teacher demonstrates this activity. When you make the product using these steps, record any observations or difficulties you have.

Step	Procedure	Observations/Difficulties
1		
2		
3		
4		
5		
6		
7		

Team and Job Assignments

List the members of your team and the job each person was assigned.

Name	Assignment
1. ______	______
2. ______	______
3. ______	______
4. ______	______
5. ______	______

Product Analysis

Analyze the tiles members of your class made by completing this section.

Productivity

A. Number of tiles the class produced ______

B. Number of workers (class members) ______

C. Productivity of the workforce (output divided by workers—A/B) ______

Quality

Write a statement describing each factor.

A. Uniformity:

- ➤ Thickness ______
- ➤ Shape ______

B. Surface smoothness ______

C. Grooves:

- ➤ Position of the tile ______
- ➤ Size (width and depth) ______

Activity Summary

How did the productivity (products produced per person) of Activity 1 compare to the productivity of Activity 2?

___

___

___

Compare the quality of the products produced in Activity 1 with those produced in Activity 2.

Uniformity of shape and thickness:

___

___

___

Surface smoothness:

___

___

___

Groove location and size:

___

___

Name ______________________ Date ______________________

Period ______________________ Score ______________________

Activity 1-2
Marble Counter

Many children have become interested in collecting and using marbles. You see an opportunity to make money on this new interest. After doing some research, you find out where you can buy marbles in large quantities. You need a way to accurately count 20 marbles, however, so they can be placed in bags. On the grid below, sketch a solution to this problem.

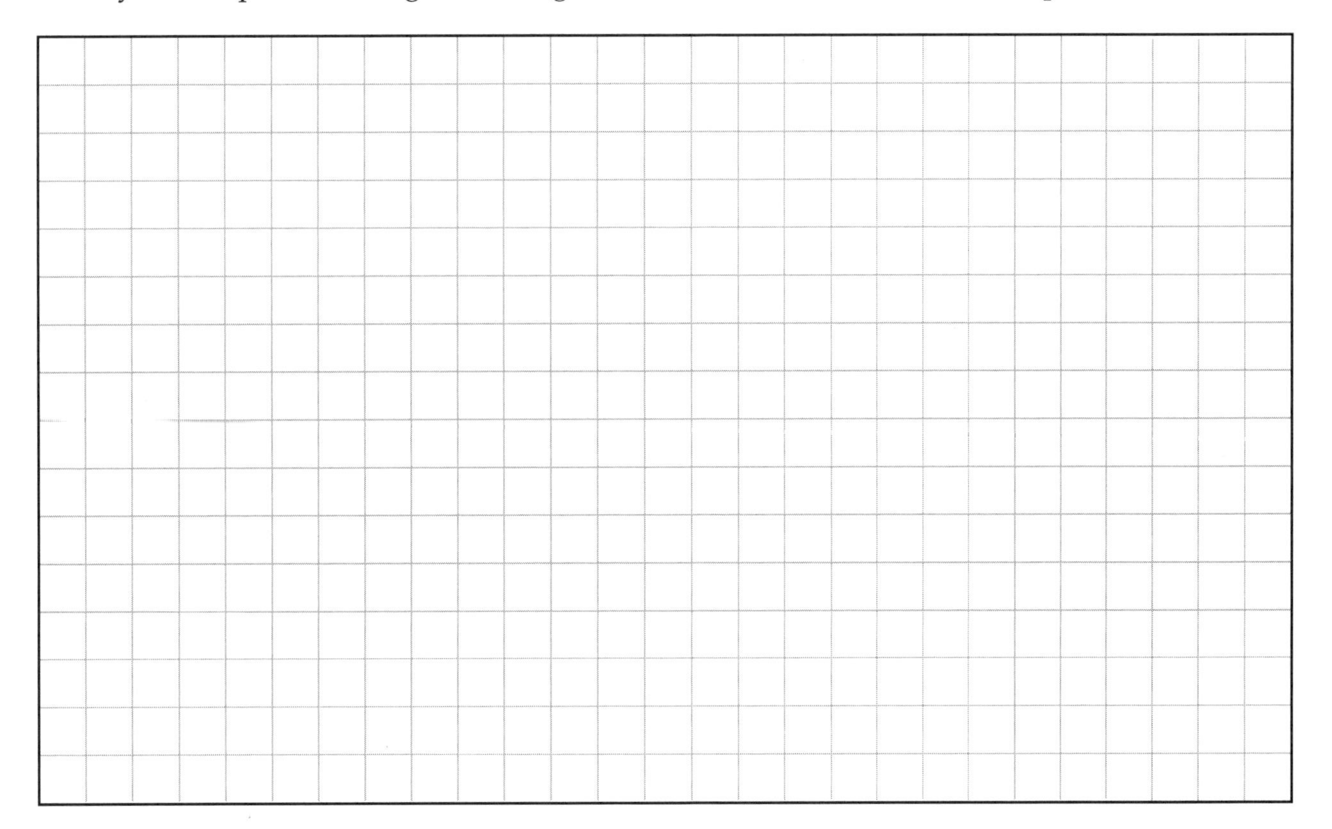

Name ______________________ Date ______________________

Period ______________________ Score ______________________

Activity 1-3
Technological Systems

For each of the following items, list its inputs, processes, outputs, and feedback.

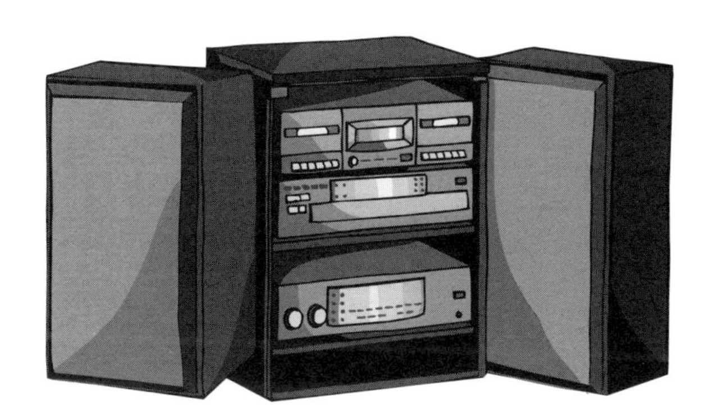

Inputs ______________________

Processes ______________________

Outputs ______________________

Feedback ______________________

Inputs ______________________

Processes ______________________

Outputs ______________________

Feedback ______________________

Inputs ____

Processes ____

Outputs ____

Feedback ____

Inputs ____

Processes ____

Outputs ____

Feedback ____

Name ______________________ Date ______________________

Period ______________________ Score ______________________

Chapter 1
Study Questions

What Is Technology?

Textbook pages: 22–33

1. List five items you own that would not exist without technology. ______________________

____

____

____

____

2. _____ knowledge explains the laws and principles governing the universe. _____ knowledge is the knowledge of the human-built world. ______________________

____

Match the statement on the left with the technological action on the right that it describes.

3. Making products or messages ______________________
4. Selecting products to meet a need ______________________
5. Measuring the impacts of using a device ______________________
6. Creating devices or systems ______________________

A. Designing
B. Producing
C. Using
D. Assessing

Name ______________________ Date ______________________

Period ______________________ Score ______________________

Chapter 2
Study Questions

Technology As a System

Textbook pages: 34–53

1. A group of parts working together to complete a task is called a(n)_________. ______________________

Match the definition with the correct part of a technological system.

2. Reason for the system	A. Input	______________
3. Result of the system	B. Process	______________
4. Resource a system uses	C. Output	______________
5. Action taken to change inputs	D. Feedback	______________
	E. Goal	
6. Information gathered to monitor output		______________

7. List the seven inputs to technological systems.
____
____
____
____
____
____
____

8. _____ processes change inputs into outputs. _____ processes plan, organize, direct, and control technological actions. ______________________
____

9. Scrap and pollution are examples of _____ outputs. ______________________

Name ______________________ Date ______________________

Period ______________________ Score ______________________

Chapter 3
Study Questions

Contexts of Technology

Textbook pages: 54–73

Name the technological context each of the following statements describes.

1. Developing and using devices and systems to plant, grow, and harvest crops. ______________________
2. Developing and using devices and systems to gather, process, and share information and ideas. ______________________
3. Using systems and processes to erect structures on the site where they will be used. ______________________
4. Developing and using devices and systems to convert, transmit, and use energy. ______________________
5. Developing and using systems and processes to convert materials into products in a factory. ______________________
6. Developing and using devices and systems promoting health and curing illness. ______________________
7. Developing and using devices and systems to move people and cargo from an origin point to a destination. ______________________

Name ______________________ Date ______________________
Period ______________________ Score ______________________

Activity 2-1
Tools and Materials As Resources

Safety

In the left column of this form, as your teacher demonstrates this activity, list the steps it will take to make the product. Record any safety procedures in the right column.

Step	Safety Precaution
1.	
2.	
3.	
4.	
5.	
6.	
7.	
8.	
9.	
10.	
11.	
12.	
13.	
14.	
15.	
16.	
17.	
18.	
19.	

(continued)

Step	Safety Precaution
20.	
21.	
22.	
23.	
24.	
25.	
26.	
27.	
28.	
29.	
30.	
31.	
32.	
33.	
34.	
35.	
36.	
37.	
38.	
39.	
40.	
41.	
42.	

Record five steps using tools necessary to produce your product. List the tools. Mark each of the tools using the following codes:

M = measuring tool
C = cutting tool
Sa = sawing tool
Sl = slicing tool
Sh = shearing tool
D = drilling tool
G = gripping tool
H = holding tool
T = turning tool
Pd = pounding tool
Po = polishing tool

Step	Tools Used	Code

Name ______________________________ Date ______________________

Period ______________________________ Score ______________________

Activity 2-2
Tools and Energy As Resources

Rubber Band Car

The following drawings show a product (rubber band-powered vehicle). You may build it, as you study about the common tools needed to make and use technological systems.

Working with your teacher, you should complete the following steps:

1. Develop a bill of materials (list of supplies) needed to make the product.
2. Prepare a procedure sheet (list of steps) for building the product. *Note:* List important safety rules with the procedure.

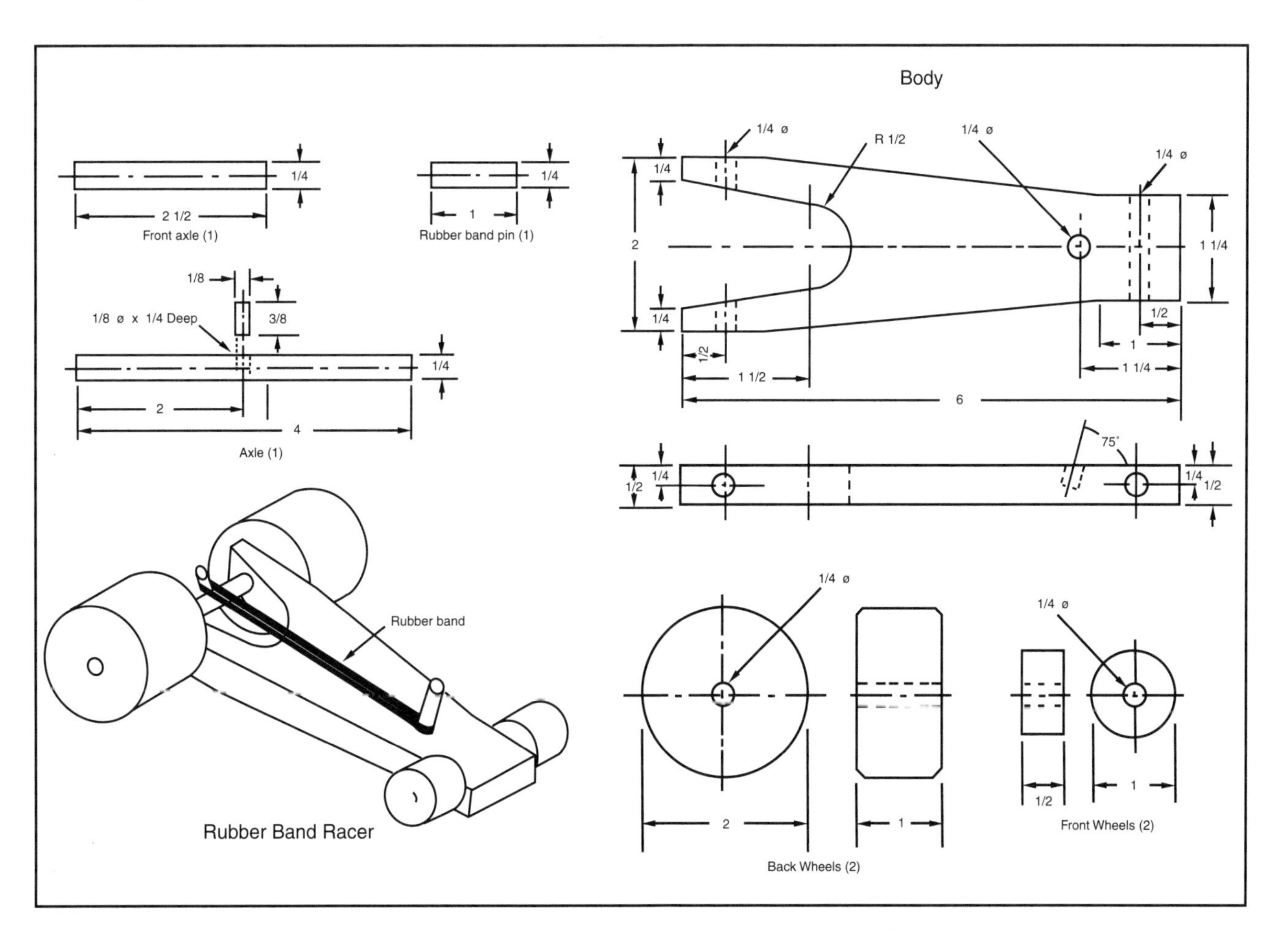

Name ______________________ Date ______________________

Period ______________________ Score ______________________

Activity 2-3
Procedure and Tool Use

Building your product requires tools. They help you extend your ability to do work. On the chart below, select five steps you have completed. List the tools you used and indicate their types:
M = measuring, C = cutting, D = drilling, G = gripping, Pd = pounding, and Po = polishing.

Step	Tools Used	Type
1.	a.	
	b.	
	c.	
	d.	
2.	a.	
	b.	
	c.	
	d.	
3.	a.	
	b.	
	c.	
	d.	
4.	a.	
	b.	
	c.	
	d.	
5.	a.	
	b.	
	c.	
	d.	

Name ______________________ Date ______________________

Period ______________________ Score ______________________

Activity 2-4
Materials As a Resource

Material Properties

Group Number: ______________________

Members: (a) ______________________ (b) ______________________

(c) ______________________ (d) ______________________

After your group has tested the materials given to you, describe their properties on the chart below. Write a brief statement describing the properties listed for each specimen.

Specimen Number	1	2	3	4	5	6
Name of Material						
Type of Material						
Density						
Hardness						
Light Reflectivity						
Torsion Strength						

Applications

You tested six specimens during this activity. Based on your test results, describe each material and give three applications for it.

Description	Applications
Specimen #1	1. 2. 3.
Specimen #2	1. 2. 3.
Specimen #3	1. 2. 3.
Specimen #4	1. 2. 3.
Specimen #5	1. 2. 3.
Specimen #6	1. 2. 3.

Name ______________________ Date ______________________

Period ______________________ Score ______________________

Activity 2-5
Product Analysis

Assignment

1. Select any product in the technology education laboratory.
2. List four materials used to make the product.
3. On the chart below, list the materials and their uses.
4. Indicate on the chart why you think the product designer chose each material used.

Product Chosen: ______________________	
1. Material used: ____________ Where it was used: ____________	Why you think it was chosen:
2. Material used: ____________ Where it was used: ____________	Why you think it was chosen:
3. Material used: ____________ Where it was used: ____________	Why you think it was chosen:
4. Material used: ____________ Where it was used: ____________	Why you think it was chosen:

Name ______________________________ Date ______________________

Period ______________________________ Score ______________________

Activity 2-6
Energy As a Resource

Motor Schematic

Draw a schematic for your motor on the grid below.

List below three ways you can improve the efficiency of your motor:

Name ______________________________ Date ______________________

Period ______________________________ Score ______________________

Activity 2-7
Energy As a Resource

Wind-Powered Generator

Given a small, low voltage, direct current motor, design a working wind-powered electric generating system. Your design will be compared with other designs produced in your class to determine the most efficient system. Use the next page for your final sketch.

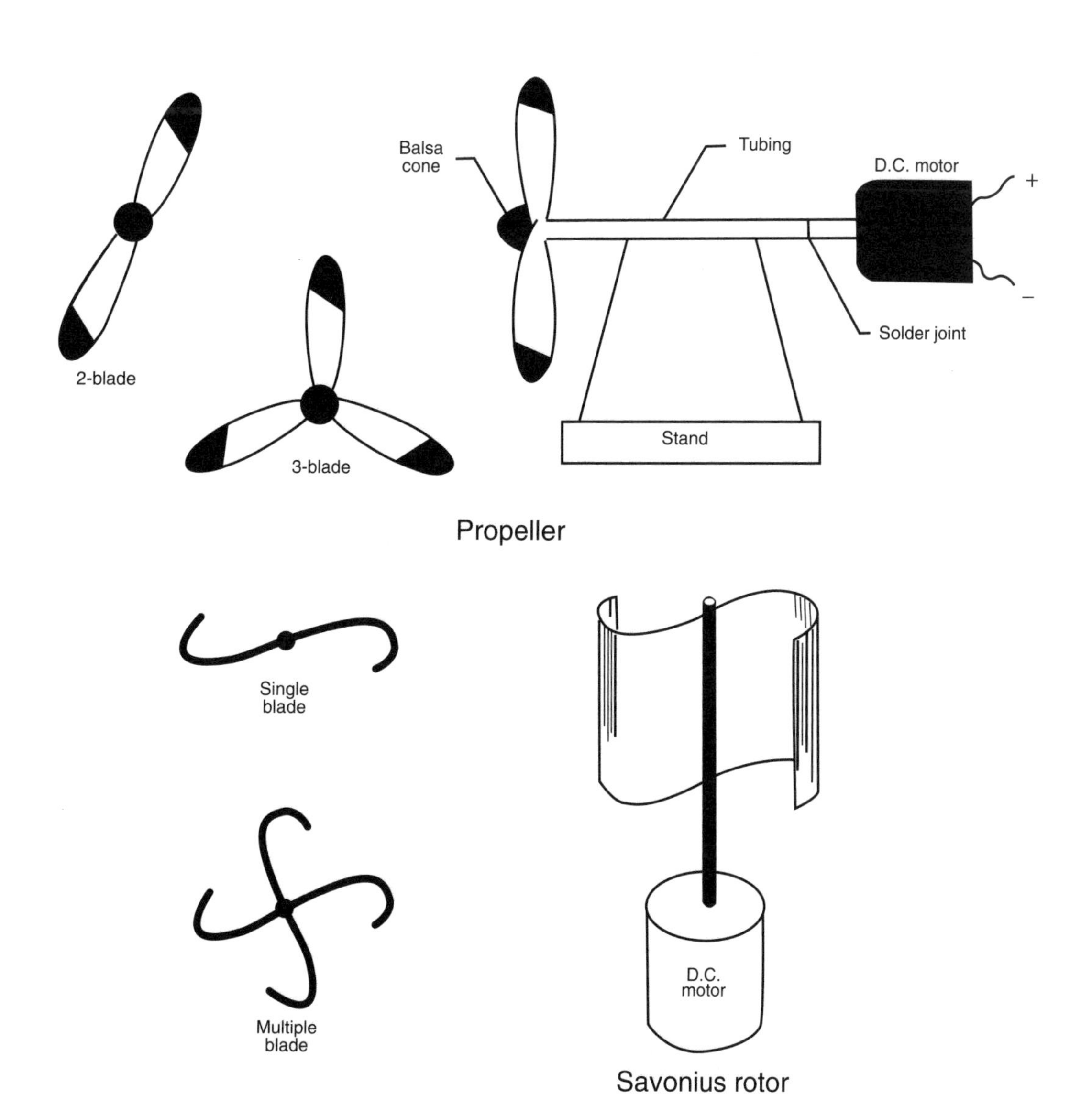

Two Examples for Designs

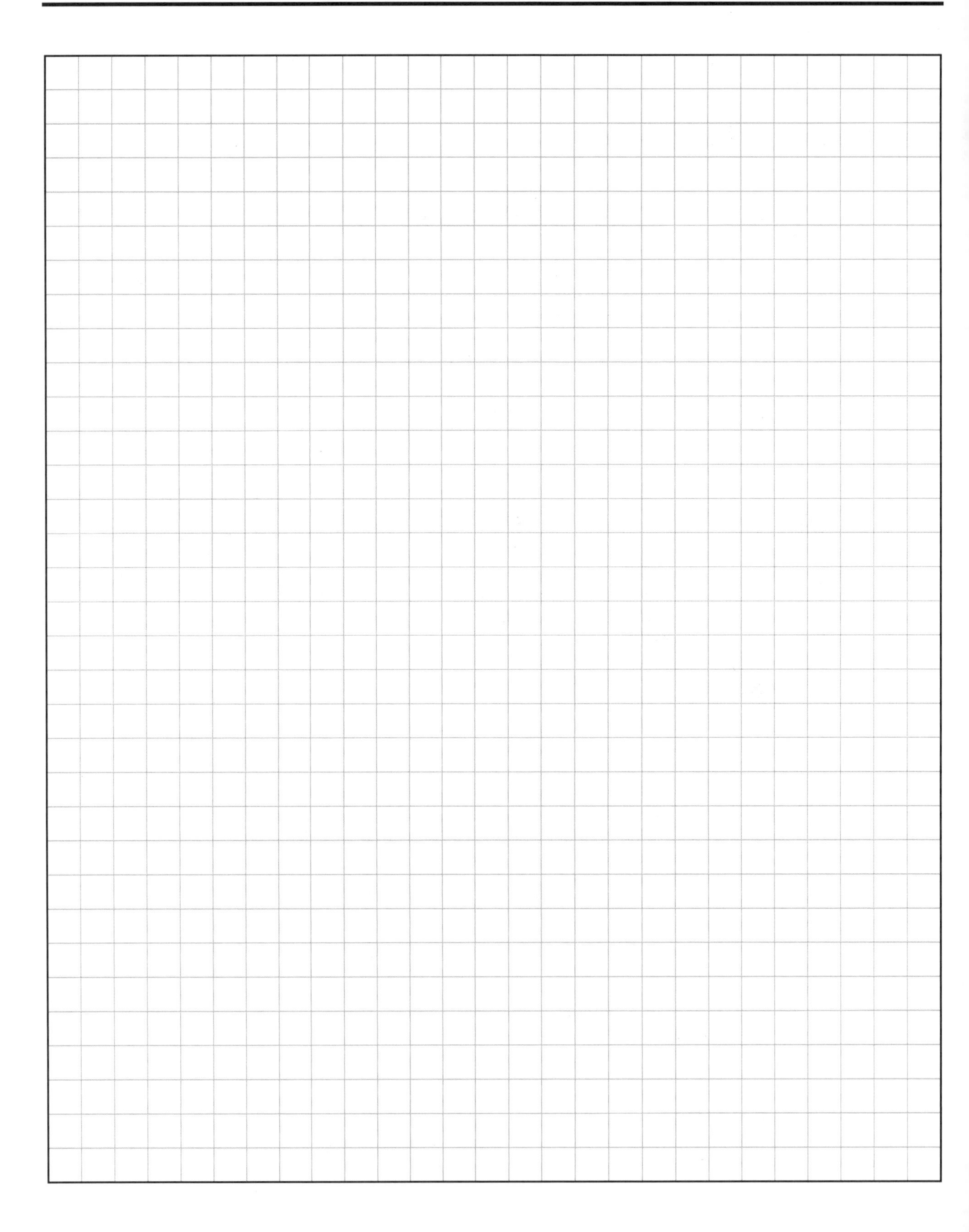

Name ______________________ Date ______________

Period ______________________ Score ______________

Activity 2-8
Information As a Resource

Note Taking

Take notes from the brief articles on pages 205–209 of your text in the space below.

Prepare an outline for your report below.

Prepare sketches of any illustrations or charts you will use in your report below.

Name ______________________ Date ______________________

Period ______________________ Score ______________________

Activity 2-9
Developing an Information Chart

The Tools and Materials As Resources (tic-tac-toe game) activity on pages 186–192 of the textbook provides you with information you need to make the product. On the chart below, list the types of information you would need to build the product and where that information was given. (D = drawings, B = bill of materials, P = procedures)

Information Needed	Where Available	Sample of the Information (copy of typical entry)
1.		
2.		
3.		
4.		

Name ______________________ Date ______________________

Period ______________________ Score ______________________

Chapter 4
Study Questions

Tools and Technology

Textbook pages: 80–105

1. What are primary tools, and why are they necessary?

____

____

____

Matching Test

Match the descriptions with the proper list of tools.

2. Tools for communicating.
3. Tools used in business and commerce.
4. Tools used for moving materials.
5. Tools used for measuring.
6. Tools used for doing math.

A. Scales, rulers, and wristwatches. ____________

B. Hand trucks, dollies, trains, and wheelbarrows. ____________

C. Calculators, computers, and abaci. ____________

D. Computers, money, and desks. ____________

E. Telephones, typewriters, computers, and pencils. ____________

7. A scissors is an example of a(n) _____-_____ lever. ____________

____

8. Properly label the parts of the first-class lever shown below:

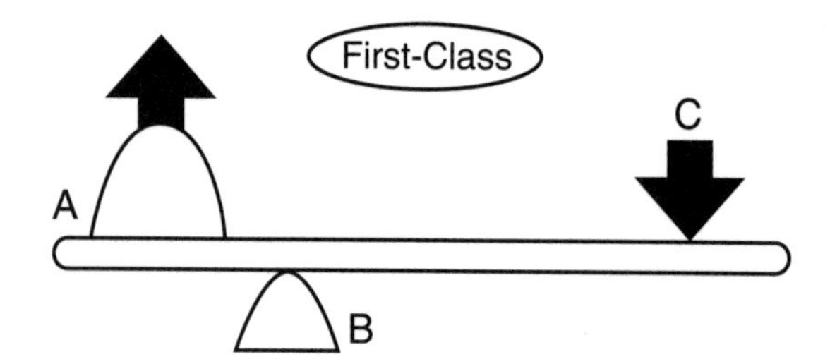

A ____________________

B ____________________

C ____________________

9. A person pedaling a bicycle is using the rear wheel as a: ____________________

   A. Force multiplier.

   B. Distance multiplier.

10. What are the six major classes of machine tools? ____________________

____

____

____

____

____

Name ______________________ Date ______________________

Period ______________________ Score ______________________

Chapter 5
Study Questions

Materials and Technology

Textbook pages: 106–127

1. Technological systems use: ______________________

   A. Two types of materials.

   B. Three types of materials.

   C. Four types of materials.

   D. No materials.

2. Products, such as chairs, have set forms or shapes. Therefore, they must be made from _____ _____. ______________________ ______________________

3. What do metals, ceramics, polymers, and composites have in common?

____

____

____

4. A basic name for materials that have never been alive is: ______________________

   A. Ceramics.

   B. Metals.

   C. Inorganic materials.

   D. Engineering materials.

5. Glass belongs to a class of engineering materials known as: ______________________

   A. Composites.

   B. Crystals.

   C. Ceramics.

   D. Plastics.

6. Rubber and plastics belong to the same general class of materials known as polymers. True or false. ______________________

7. Indicate whether the following materials are exhaustible or renewable.

   A. Wood. ____________________

   B. Rubber. ____________________

   C. Animals. ____________________

   D. Corn. ____________________

   E. Iron. ____________________

   F. Coal. ____________________

   G. Soil. ____________________

8. Name two methods of extraction used to get exhaustible materials from the earth.

   ____

   ____

   ____

9. Crushing rocks and grinding grain are examples of _____ processing. ____________________

10. Using heat to process materials is called _____ processing. ____________________

11. What are the seven major properties of materials?

   ____

   ____

   ____

   ____

   ____

   ____

   ____

12. How a material reacts to forces or conditions is called its _____ properties. ____________________

Name ______________________________ Date ______________________

Period ______________________________ Score ______________________

Chapter 6
Study Questions

Energy and Technology

Textbook pages: 128–149

1. Energy is used in natural and technological systems. True or false. ______________________

2. Energy is the _____ to do _____. ______________________
____

3. Stored energy is called _____ energy. ______________________

4. Energy in motion is called _____ energy. ______________________

5. Power and energy are (select all correct answers): ______________________
   A. Different things. Power is force times distance moved, while energy is the ability to do work.
   B. Different things. Energy is the ability to do work, while power is the speed at which work is done.
   C. The same thing.

6. The amount of energy required to raise one pound of water 1°F, at sea level, is called a(n) _____. ______________________

7. Name the six forms of energy. ______________________
____
____
____
____
____

8. Coal, petroleum, and natural gas are examples of _____ fuels. ______________________

9. Splitting atoms to release energy is called nuclear _____. _______________

10. Energy from the natural heat below Earth's crust is called _____ energy. _______________

11. Energy from organic matter is called _____. _______________

Name ______________________ Date ______________________

Period ______________________ Score ______________________

Chapter 7
Study Questions

Information and Technology

Textbook pages: 150–165

1. Name each of the following types of knowledge:
   A. Knowledge of how people have expressed themselves through arts, developed values, and established societies. ______________________
   B. Knowledge of how people use tools and materials to make things. ______________________
   C. Knowledge of the laws governing the universe. ______________________
   D. Knowledge of how to use words and numbers to describe objects and events. ______________________
   E. Knowledge about personal, society, and religious values. ______________________
   F. Knowledge about making and using things. ______________________
2. Individual facts, statistics, and ideas are called _____. ______________________
3. Organized and sorted data is called _____. ______________________
4. Information humans can apply to situations is called _____. ______________________
5. Designers only use technological knowledge to solve problems. True or false. ______________________
6. Name the three basic types of research. ______________________

   ____

   ____
7. Research describing how other people solved a problem is called _____ research. ______________________

8. Research that gathers information by measuring and describing items or events is called _____ research. ______________________

9. Research that changes situations and compares the results is called _____ research. ______________________

Name ______________________ Date ______________________

Period ______________________ Score ______________________

Chapter 8
Study Questions

People, Time, Money, and Technology

Textbook pages: 166–185

1. People who are paid a salary are often called _____-collar workers. ______________________
2. The three major groups hourly (blue-collar) workers can be separated into are: ______________________ ______________________ ______________________
3. A company is a(n) _____ institution that uses _____ to produce a(n) _____, structure, or _____ with the intent of making a(n) _____. ______________________ ______________________ ______________________ ______________________ ______________________
4. People who organize resources to produce products efficiently are called _____. ______________________
5. _____ design products, processes, and structures. ______________________
6. People who work closely with engineers to implement their work are called _____. ______________________
7. Moving up the _____ means getting better jobs as your skills and knowledge improve. ______________________
8. _____ is the unit of measure for a duration of an event. ______________________
9. _____ is used to purchase _____ used to _____ a product or service exchanged in the _____ into _____. ______________________ ______________________ ______________________ ______________________ ______________________

Name ______________________ Date ______________________

Period ______________________ Score ______________________

Activity 3-1
Shading and Rendering

Using colored pencils, shade and render the bookcase below. Use the Sun icon as the light source.

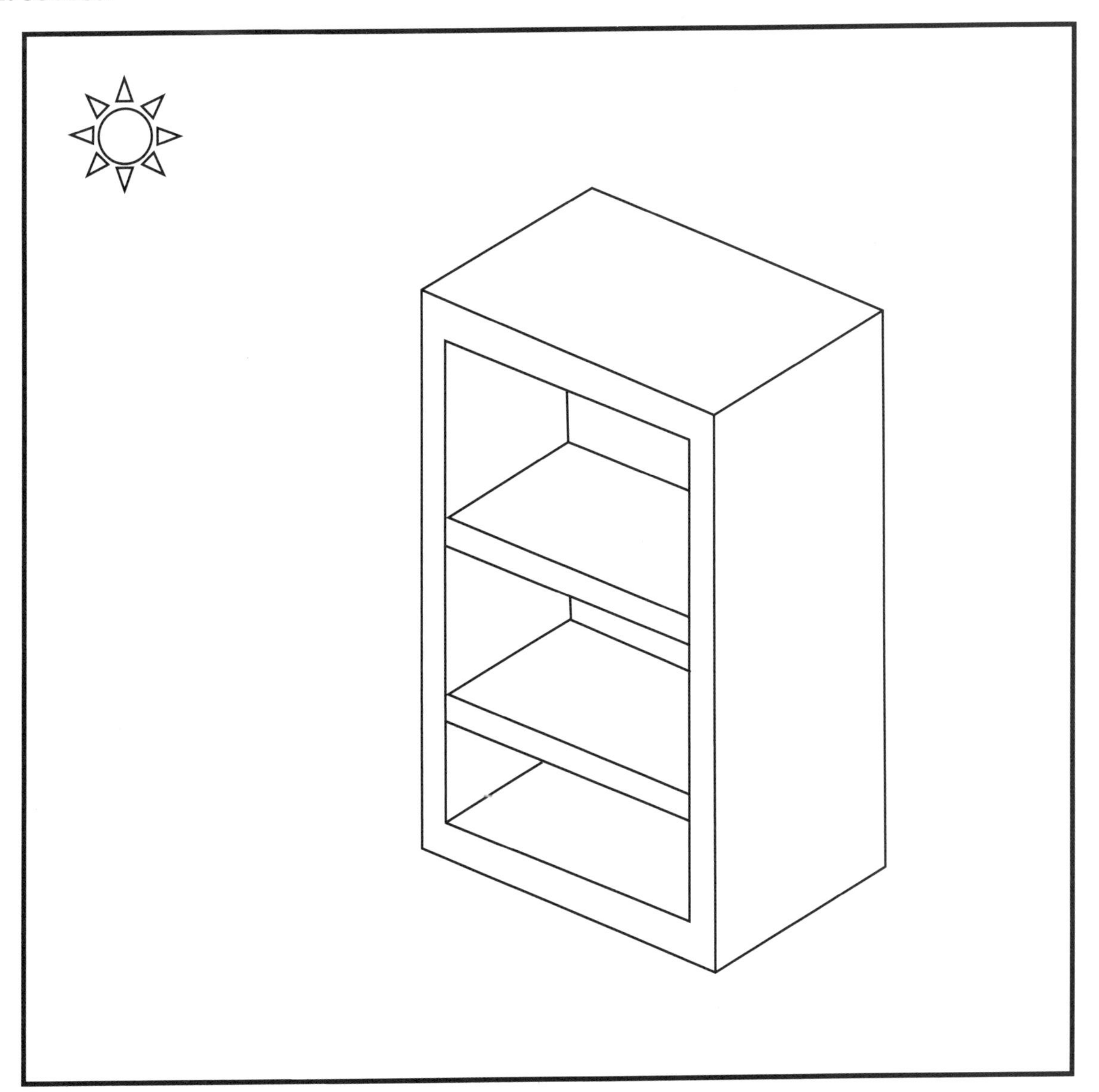

Name ______________________ Date ______________________

Period ______________________ Score ______________________

Activity 3-2
Orthographic Sketching

Sketch the front, top, and right side views of the three objects below.

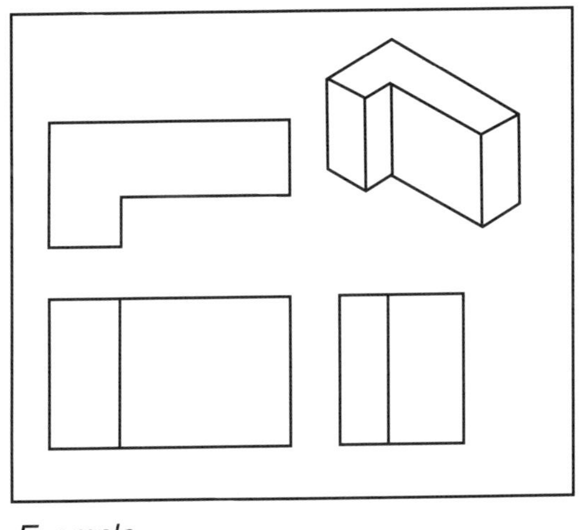

Example

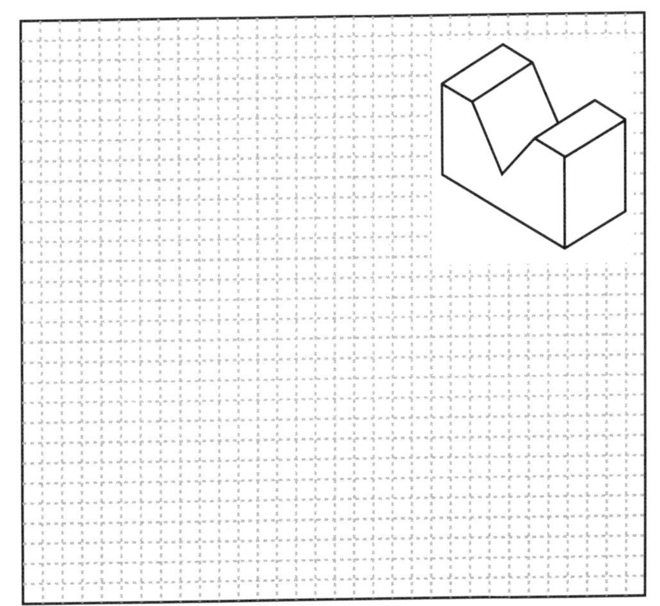

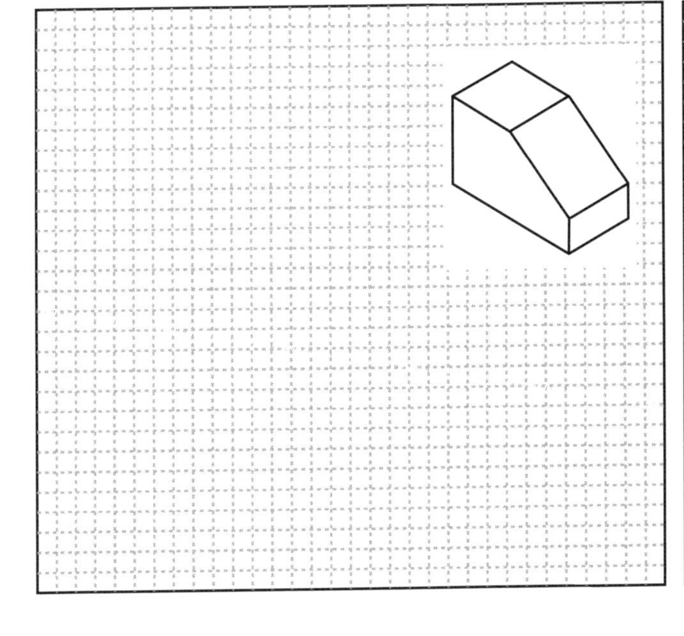

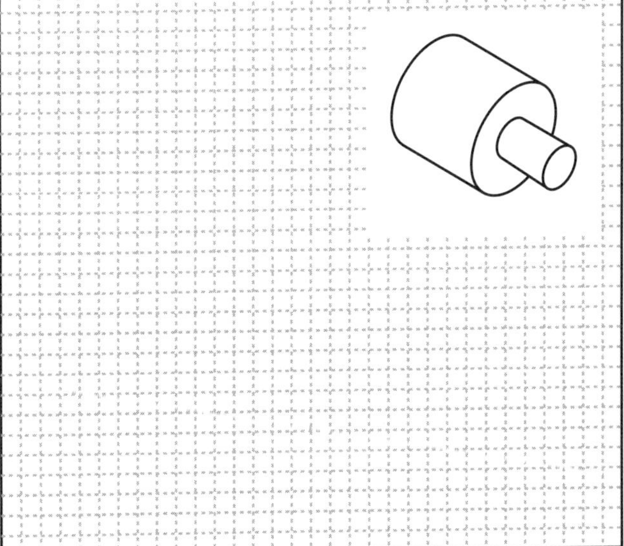

Name ______________________________ Date ______________________

Period ______________________________ Score ______________________

Activity 3-3
Communication Design

Design Brief

Complete the following sheet to create a design brief to guide the design of your commercial.

Statement of Problem

The technology education department needs a promotional video to increase student enrollment. It especially wants to focus on the course in which you are enrolled.

Additional Problems:

____

____

____

Criteria

The solution must be the following:

1. Informational.
2. Student designed.
3. 30 seconds in length.
4.
5.
6.
7.

Constraints

The solution cannot have any of the following characteristics:

1. Be produced outside of the school.
2. Contain illegal or immoral activities.
3.
4.
5.
6.
7.

Storyboard

Use this storyboard to create a solution to the video.

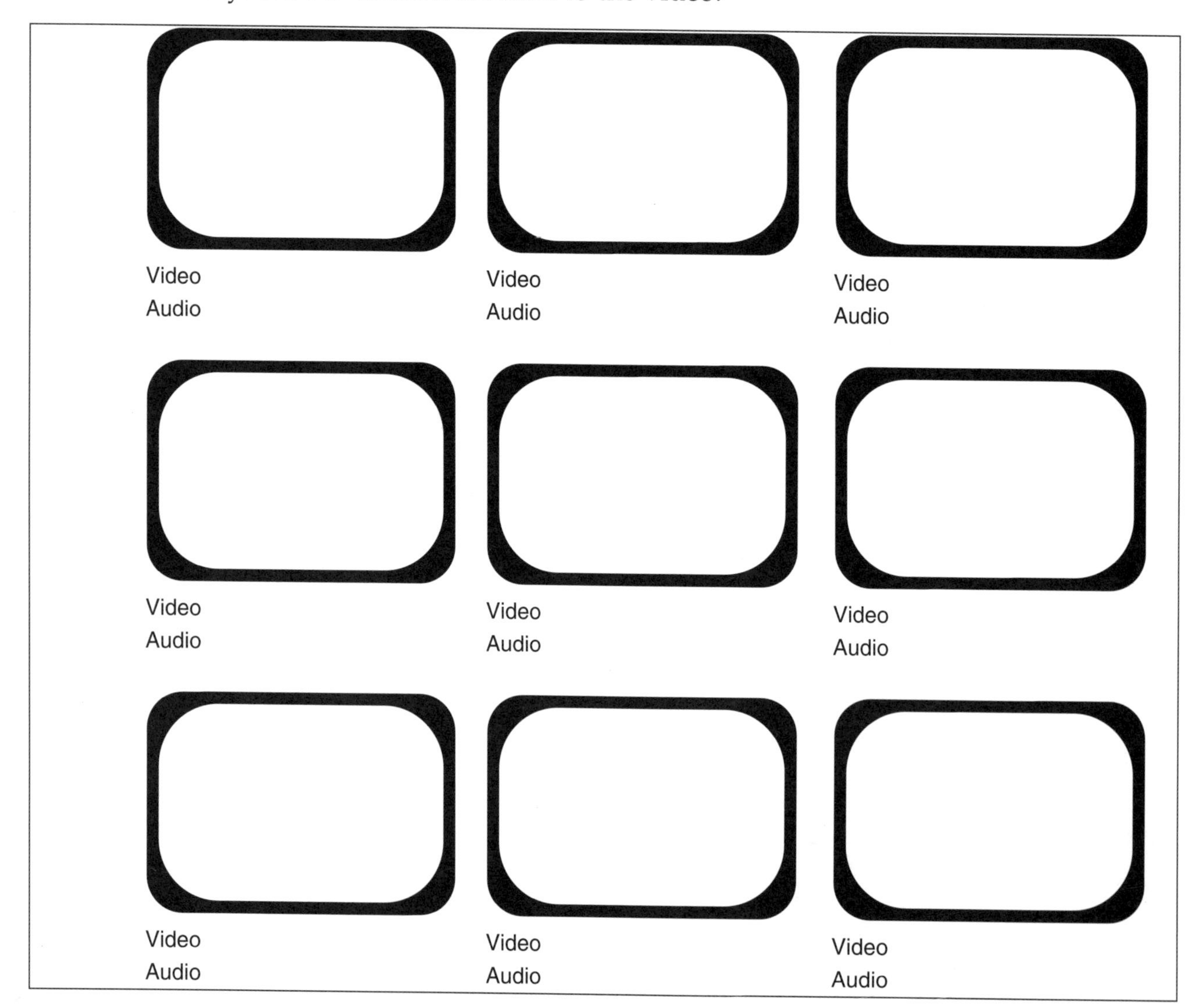

Script

Title of video: ______________________________

Actor	Text to be spoken

Page ____ of ________

Improving Designs

After viewing other design solutions, make suggestions on ways you can improve your own solution.

Improvement 1	
What changes would you make?	Why is this change needed?

Improvement 2	
What changes would you make?	Why is this change needed?

Improvement 3	
What changes would you make?	Why is this change needed?

Name ______________________ Date ______________________

Period ______________________ Score ______________________

Activity 3-4
Locker Design

Situation

A company specializing in the design of organizers has hired you. It designs CD holders, VCR tape holders, desk organizers, as well as pen and pencil holders. The company is ready to branch out into a new market. The managers have hired you to be the head designer of their new locker organizer. They feel you are the best person for the job because you know the likes and dislikes of students. Your managers also know you have to deal with trying to organize your own locker every day.

Design Problem

The company realizes many schools have different sized lockers, so it wants you to create an organizer to fit your own locker, and it will scale this organizer to fit different sized lockers. You have the choice of the type of organizer you will create. This organizer can sit on the locker shelf, hang from the door, or fill the entire locker. The company's only requirement is that the organizer must help students organize the items they store inside their lockers.

The company wants several pieces of the design from you. It requires the following items:

1. Design sketches.
2. A refined, colored sketch.
3. A dimensioned sketch.
4. A model.

Name ______________________________ Date ______________________

Period ______________________________ Score ______________________

Activity 3-5
Bio-Related Feeder

Challenge

Your challenge is to design a device that will provide water or food, on a regular schedule, to a plant or animal. You will begin by selecting the exact function of your device. Then, you will make sketches and create a display of your design.

Design Problem

Your device will supply: ☐ Water ☐ Food

The food or water is for a: ☐ Plant ☐ Animal

Note—Specify type or species: ______________________________

Your device will distribute the supply: ☐ Once a day ☐ Twice a day ☐ Every four hours ☐ Once an hour ☐ Other: ______________

Design Sketches

Sketch possible designs for the device below.

Name ______________________ Date ______________________

Period ______________________ Score ______________________

Chapter 9
Study Questions

Invention and Innovation

Textbook pages: 212–231

1. An important time for invention occurred between 1750 and 1850 and was called the ________. ______________________
2. Which of the following is a discovery? ______________________
   A. Telephone.
   B. Gravity.
   C. Manufacturing line.
   D. Bronze.
3. Name the three categories of inventions. ______________________
   ____
   ____
4. A process beginning with a problem and ending with a solution is known as a(n) ________ process. ______________________
5. List three characteristics of an inventor. ______________________
   ____
   ____
6. An invention that makes our lives better is known as a(n) ________ invention. ______________________
7. An invention created for the fun of inventing is known as a(n) ________ invention. ______________________

Matching Test

Match the descriptions with the proper term.

8. New, unique device. __________
9. Old device for a new purpose. __________
10. New change to an old device. __________

A. Adaptation.
B. Innovation.
C. Invention.

Name ______________________ Date ______________

Period ______________________ Score ______________

Chapter 10
Study Questions

The Design Process

Textbook pages: 232–253

1. __________ is the act of planning used to solve a problem. ______________
2. List three designed products. ______________

   ____

   ____

Matching Test

Match the descriptions with the proper term.

3. Topic or emotion.	A. Form.	______________
4. Look of a solution.	B. Function	______________
5. How well the solution works.	C. Content	______________

6. Artistic and engineering design solve the same types of problems. True or false. ______________
7. Organizing parts to solve a problem is an example of __________ design. ______________
8. The creation of devices meeting needs or wants is an example of __________ design. ______________

9. List the eight steps of the design process.

Name ______________________________ Date ______________________

Period ______________________________ Score ______________________

Chapter 11
Study Questions

Identifying Problems

Textbook pages: 254–267

1. A human __________ is a requirement to live. ______________________

Matching Test

Match the descriptions with the types of problems. (The answers may be used more than once.)

2. Automobile.	A. Need.	______________________
3. Food.	B. Want.	______________________
4. House.		______________________
5. DVD player.		______________________

6. Problems dealing with a large group of people are __________ problems. ______________________

7. Academic problems are a type of __________ problem. ______________________

8. A problem statement should state the exact solution. True or false. ______________________

9. The __________ are the elements of the problem that must be solved. ______________________

10. List the four types of constraints. ______________________

____

____

____

11. A(n) __________ lists the problem, criteria, and constraints. ______________________

Name ______________________________ Date ______________________

Period ______________________________ Score ______________________

Chapter 12
Study Questions

Researching Problems

Textbook pages: 268–283

1. List the three major types of research. ______________________

____

____

2. A group of people using certain products are called __________. ______________________
3. The research process is similar to the __________. ______________________
4. Raw facts and figures are known as __________. ______________________
5. Aesthetic features relate to the: ______________________
   A. Function of the product.
   B. Cost of the product.
   C. Look of the product.
   D. Safety of the product.
6. Analyzed data is called __________. ______________________

Matching Test

Match the image with the type of graph.

7. 

8. 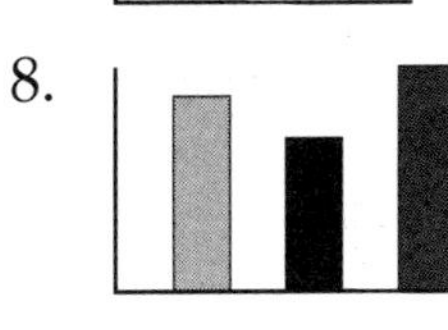

9. 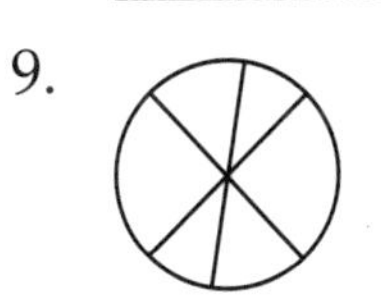

A. Line graph. ______________________

B. Pie graph. ______________________

C. Bar graph. ______________________

Name ______________________ Date ______________

Period ______________________ Score ______________

Chapter 13
Study Questions

Creating Solutions

Textbook pages: 284–309

1. __________ is the process of creating new ideas to solve a problem. ______________

Matching Test

Match the descriptions with the correct method of creating ideas.

2. A group process in which people list as many ideas as possible. ______________
3. A process in which you use a diagram to organize your thoughts. ______________
4. A process in which the designers ask why things are done a certain way. ______________

A. Brainstorming.

B. Questioning.

C. Using a graphic organizer.

5. A(n) __________ is a quick record of what the designer is thinking. ______________
6. Pictorial drawings show objects in __________ dimensions. ______________
7. List the five basic shapes in isometric sketching. ______________

____

____

____

____

8. __________ helps to show how objects look in the light. ______________
9. A(n) __________ sketch shows the front of the object "flat" on the paper and the side and top at an angle. ______________
10. Vanishing points are used in __________ sketches. ______________

Name ______________________ Date ______________________

Period ______________________ Score ______________________

Chapter 14
Study Questions

Selecting and Refining Solutions

Textbook pages: 310–329

1. When clients make the design selection, it is known as __________ selection. ______________________
2. The __________ of design are the design criteria. ______________________

Matching Test

Match the descriptions with the correct criteria.

3. The look of the solution. ______________________
4. The moving parts of the solution. ______________________
5. The comfort of the solution. ______________________
6. The manufacturing of the solution. ______________________
7. The cost of the solution. ______________________

A. Production.
B. Human factors.
C. Appearance.
D. Finances.
E. Function.

8. The __________ is the amount of light and dark in the design. ______________________
9. The surface of an object is the __________. ______________________
10. The __________ is the weight of the object on both sides. ______________________
11. A full color sketch is a(n) __________. ______________________
12. List the three types of dimensions. ______________________

____

____

Name ______________________________ Date ______________________

Period ______________________________ Score ______________________

Chapter 15
Study Questions

Modeling Solutions

Textbook pages: 330–359

1. A(n) __________ is a representation of a product, system, process, or idea. ______________________
2. List the three reasons models are used. ______________________

____

____

Matching Test

Match the example with the correct type of model.

3. Pie chart. ______________________
4. Formula. ______________________
5. Surface model. ______________________
6. Mock-up. ______________________

A. Mathematical.
B. Physical.
C. Computer.
D. Graphic.

7. A(n) __________ is a working model. ______________________
8. List the four main types of sheet materials. ______________________

____

____

____

9. Polystyrene is a type of __________. ______________________
10. __________ is a process that shapes and smoothes the model in one step. ______________________

Name ______________________ Date ______________________

Period ______________________ Score ______________________

Chapter 16
Study Questions

Testing Solutions

Textbook pages: 360–381

1. A(n) __________ is an experiment or examination. ______________________
2. A(n) __________ is a judgment made from the results of an examination. ______________________
3. List the five characteristics tested. ______________________

   ____

   ____

   ____

   ____
4. The __________ is how long a solution will work. ______________________
5. A compression test would be used to test __________. ______________________
6. A soundproof room is an example of a __________. ______________________

   A. Laboratory environment.

   B. Field test environment.

   C. Virtual environment.
7. A road course is an example of a __________. ______________________

   A. Laboratory environment.

   B. Field test environment.

   C. Virtual environment.
8. A(n) __________ test is one that can be trusted. ______________________
9. __________ is an example of a company that writes testing standards. ______________________
10. A(n) __________ is a compromise designers must face. ______________________

Name ______________________ Date ______________

Period ______________________ Score ______________

Chapter 17
Study Questions

Communicating Solutions

Textbook pages: 382–409

1. A(n) __________ is a document listing and describing the parts needed to build a product. ______________
2. __________ drawing is very neat and accurate. ______________
3. A Material Safety Data Sheet (MSDS) is a type of __________. ______________
4. __________ drawings communicate buildings and structures. ______________

Matching Test

Match the description with the correct type of drawing.

5. Shows the exact size and shape. ______________
6. Shows how parts fit together. ______________
7. Shows a system. ______________

A. Detailed.

B. Schematic.

C. Assembly.

8. __________ drawings show objects in separate two-dimensional views. ______________
9. The __________ contains different line types. ______________
10. Isometric drawings are drawn with __________ degree angles. ______________
11. An oblique drawing using 1/2 scale on the sides is called a(n) __________ oblique. ______________
12. List the three traditional drawing tools. ______________

____

Name ______________________ Date ______________________

Period ______________________ Score ______________________

Chapter 18
Study Questions

Improving Solutions

Textbook pages: 410–425

1. The money left over after the bills are paid is known as __________. ______________________
2. A piece of software fixing a problem is called a(n) __________. ______________________
3. Name a commonly recalled type of product. ______________________
4. List four sources of improvement. ______________________

   ____

   ____

   ____
5. The common size of objects, like batteries, is a result of __________. ______________________
6. Once the design process reaches the improvement step, it is finished. True or false. ______________________

Name ______________________ Date ______________________

Period ______________________ Score ______________________

Activity 4-1
Sketching and Estimating

Name of the construction project: ______________________

Part of the structure your group is producing: ______________________

Date to be completed: ______________________

On the grid below, draw a sketch of the building component you are building. Label all the major parts.

Estimating Construction Costs

Estimating the cost of a structure requires making educated guesses about how much material will be needed, how much labor is involved, and what the land will cost. *Labor costs* are estimated by multiplying the number of workers by the estimated number of hours required to complete the structure. *Equipment costs* are based on money that must be spent to rent or buy tools, machines, and other equipment. *Material costs* are arrived at by checking prices in a catalog or at a lumberyard. *Land costs* can be determined by checking real estate ads in local newspapers.

Your assignment is to determine the cost of building the shed in your textbook activity. Place your estimates in the chart. Use the equipment and supply list on page 629 of your textbook.

Construction Estimate for Storage Shed		
Item		Estimate
Land		
Labor (five crew members)		
Equipment (list all tools on page 629)		
Quantity	Description	
Materials		
Quantity	Description	

Name ______________________ Date ______________________

Period ______________________ Score ______________________

Activity 4-2
Model Bridge Design Contest

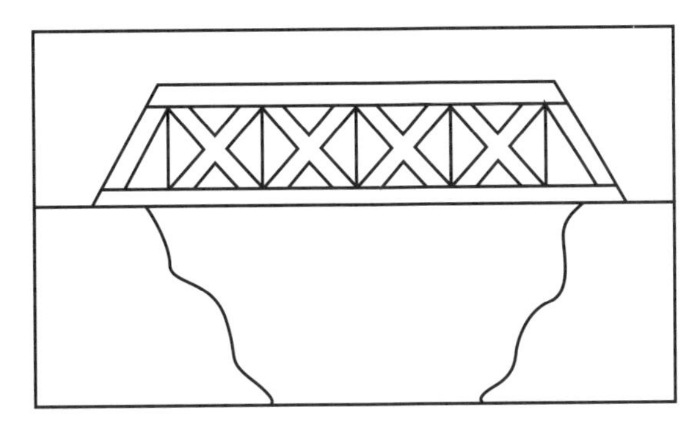

Design Brief

Your task is to design a 12" long × 2" high model bridge. You will have ten pieces of 1/4" × 1/4" × 12" material. Your model will be compared with others built in the class, but you may use any design. The bridge holding the most weight is the best design. Each bridge will be tested to its breaking strength on a test stand, similar to the one at the bottom of this page.

Procedure

1. Research bridge designs in the school library.
2. Design and draw a full-scale bridge framework.
3. Cut the parts needed to make the two sides of the bridge.
4. Glue the bridge sides together.
5. Assemble the two sides with spacers holding the sides 1 1/2" apart.
6. Test the strength of the bridge, recording the maximum weight it supported.
7. Compare the results of your bridge with those of other class members.
8. Determine the "best design" (the bridge holding the heaviest load before breaking).

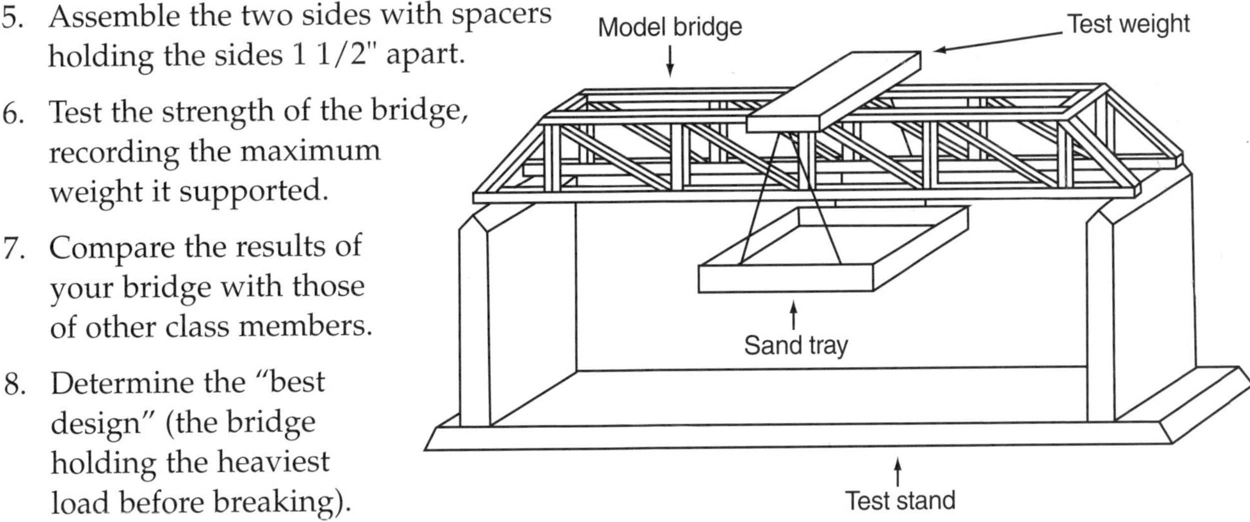

Name ______________________ Date ______________________

Period ______________________ Score ______________________

Activity 4-3
Energy Conversion Technology

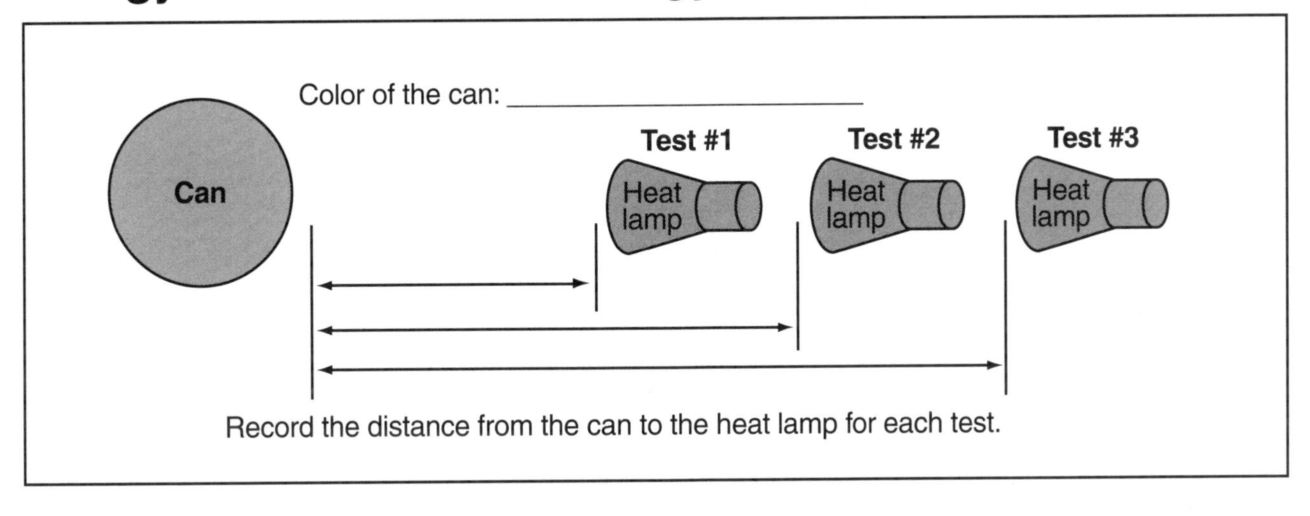

Test #1 — Distance from the lamp to the can: __________ inches.

Time	Can Color					
________ minutes						
________ minutes						
________ minutes						

Test #2 — Distance from the lamp to the can: __________ inches.

Time	Can Color					
________ minutes						
________ minutes						
________ minutes						

Test #3 — Distance from the lamp to the can: __________ inches.						
Time	Can Color					
________ minutes						
________ minutes						
________ minutes						

Summary

1. Which color absorbed the most light energy? ______________________________

   Why do you think this happened? ______________________________

2. Which color absorbed the least light energy? ______________________________

   Why do you think this happened? ______________________________

3. Which color would you use for each application given below? Give a reason for your selection.

   a. Roof of a home in the arctic. Color: __________ Why? ______________________________

   ____

   b. Heat absorbing wall in a passive solar home. Color: __________ Why? ______________

   ____

   c. Southern side of a building in Phoenix, Arizona. Color: __________ Why? __________

   ____

   d. Air-conditioned bus. Color: __________ Why? ______________________________

   ____

Name ______________________ Date ______________________

Period ______________________ Score ______________________

Activity 4-4
Information and Communication Technology

Developing a Code

You must communicate information that will allow a person to move through a maze. Your challenge is to develop a code that can be communicated on a telegraph that will tell a person to move up, down, left, or right; stop; and pass a certain opening. Enter your code in the space below.

a. Move right.	b. Move left.	c. Move up.	d. Move down.

e. Pass the first opening.	f. Pass the second opening.	g. Pass the third opening.	h. You made it!

Developing the Message

Select the maze shown below for your group. Determine the movements that will be needed to move from the starting point to the finish line. Write the code on the next page.

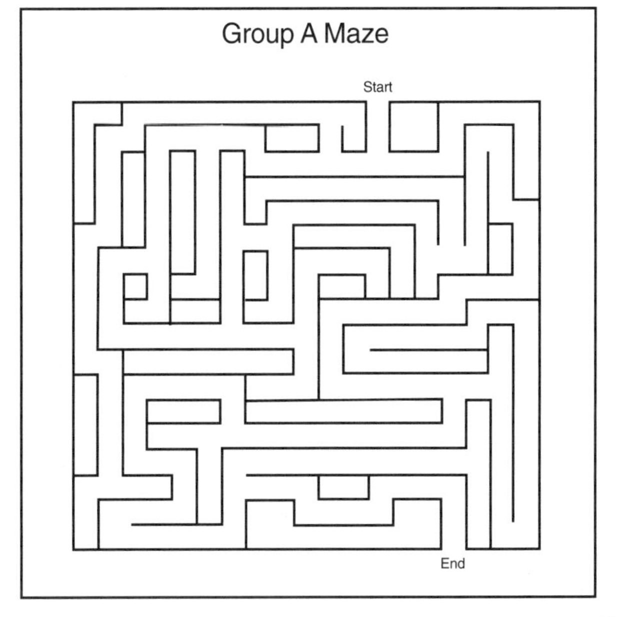

Codes for Solving the Maze

Move	Code
1.	
2.	
3.	
4.	
5.	
6.	
7.	
8.	
9.	
10.	

Move	Code
11.	
12.	
13.	
14.	
15.	
16.	
17.	
18.	
19.	
20.	

Move	Code
21.	
22.	
23.	
24.	
25.	
26.	
27.	
28.	
29.	
30.	

When you have finished this part, send the message.

Receiving the Message

Your partner will send you a message to solve your group's maze. Record the message below.

Move	Message
1.	
2.	
3.	
4.	
5.	
6.	
7.	
8.	
9.	
10.	

Move	Message
11.	
12.	
13.	
14.	
15.	
16.	
17.	
18.	
19.	
20.	

Move	Message
21.	
22.	
23.	
24.	
25.	
26.	
27.	
28.	
29.	
30.	

Assessing the System

Was your group able to receive the message? ______________________

If not, what caused the problem? ______________________

____

____

What were the inputs, processes, and outputs to the telegraph system?

Inputs: ______________________

Processes: ______________________

Outputs: ______________________

What was the distance over which the message was sent? _______

Modifying the System

How can the telegraph be modified so a person who is deaf can use it?

Working as a group, modify the design to give a visual message.

Describe how you can modify the design to provide a visual message.

Sketch the modified design on the following page.

On the grid below, sketch a modified telegraph system that can provide visual messages.

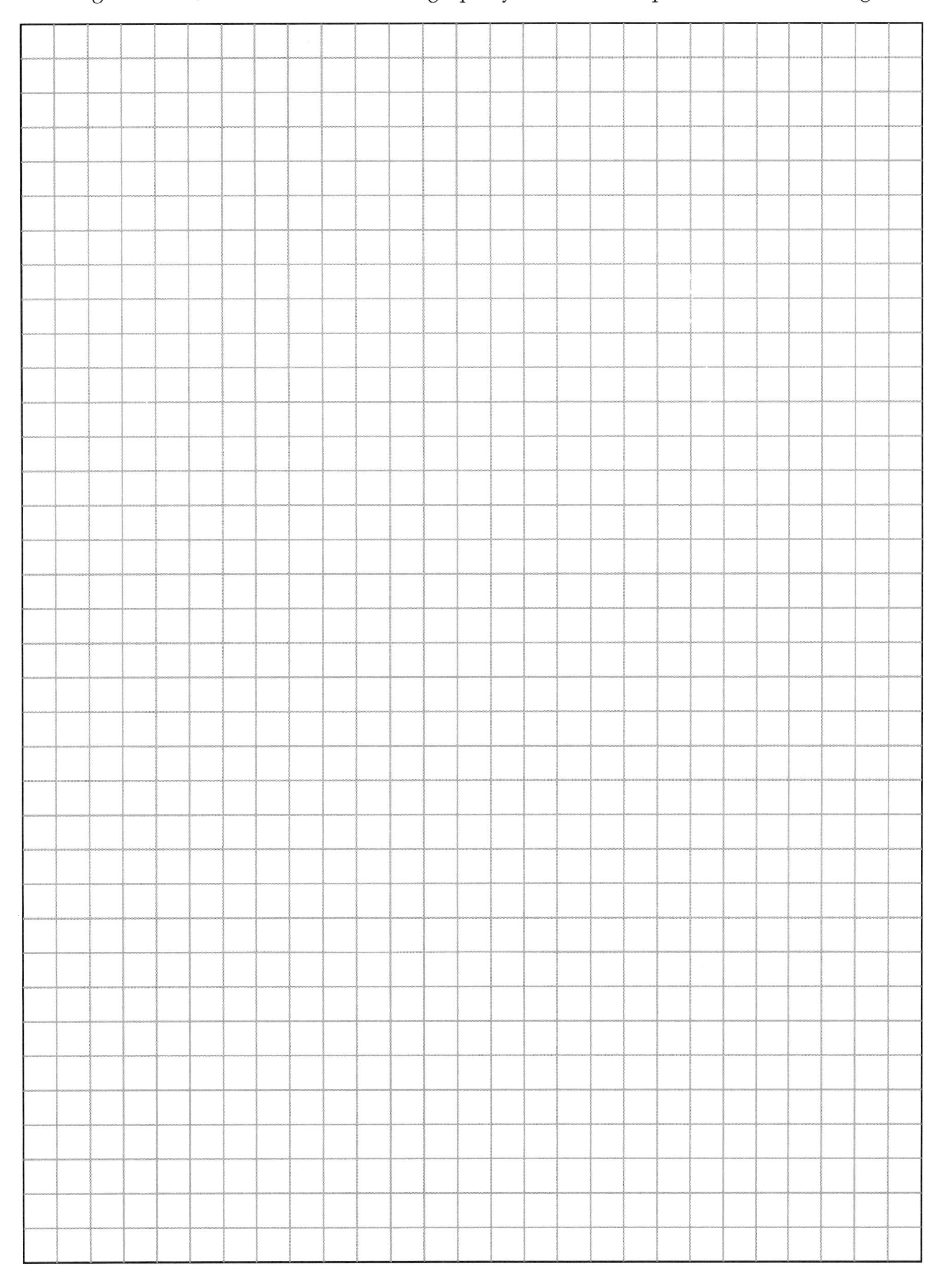

Name ______________________ Date ______________________

Period ______________________ Score ______________________

Activity 4-5
Billboard Message

Design Brief

Graphic designers must be concerned about using printed messages to attract readers. They must determine who their audience is and how to get its attention. These designers must then design an effective message. Select one of the design problems below and develop it for the audience specified.

Problem #1

Design a billboard message promoting safe bicycle riding habits among students in your school.

Problem # 2

Recently, a problem of trash on the school grounds has developed. Design a poster addressing this problem. Direct the message to the students in your school.

Message Design

Theme or slogan to be used: ______________________

Message layout:

Name ______________________ Date ______________________

Period ______________________ Score ______________________

Activity 4-6
Process Demonstration

During the demonstration of the processes to be used to manufacture your product, complete all three columns of this chart. In the first column, write a very brief description of each step. In the second column, list the tools or machines used. In the third column, list any safety precautions you should observe.

Step	Tool or Machine	Safety Consideration
1.		
2.		
3.		
4.		
5.		
6.		
7.		
8.		
9.		
10.		
11.		
12.		

Step	Tool or Machine	Safety Consideration
13.		
14.		
15.		
16.		
17.		
18.		
19.		
20.		
21.		
22.		
23.		
24.		
25.		
26.		
27.		
28.		
29.		
30.		

Name ______________________________ Date ______________________

Period ______________________________ Score ______________________

Activity 4-7
Recipe Holder

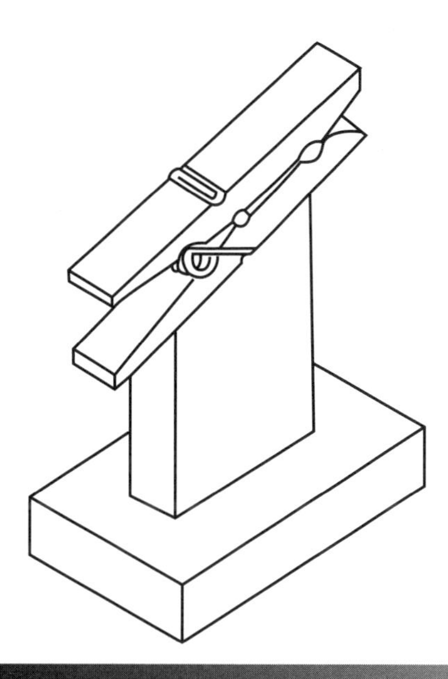

Assignment

Study the drawings for the recipe holder. Then, draw up a bill of materials for this holder. On the following page, list the procedure for making the base and upright.

Bill of Materials				
Quantity	Part Name	Size		
		Thickness	Width	Length

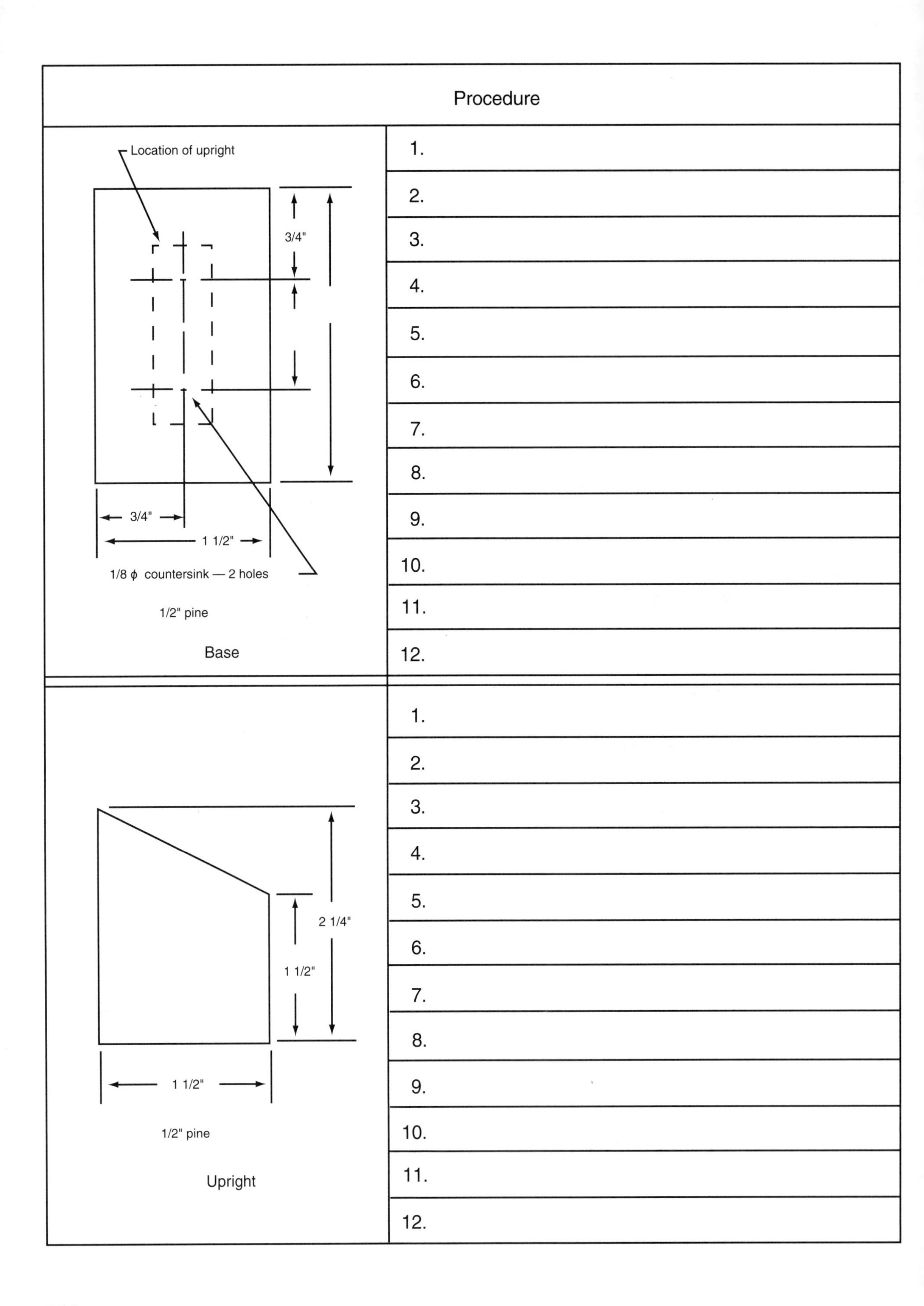
Procedure
Location of upright
3/4"
3/4"
1 1/2"
1/8 ϕ countersink — 2 holes
1/2" pine
Base
1.
2.
3.
4.
5.
6.
7.
8.
9.
10.
11.
12.
2 1/4"
1 1/2"
1 1/2"
1/2" pine
Upright
1.
2.
3.
4.
5.
6.
7.
8.
9.
10.
11.
12.

Name ______________________ Date ______________________

Period ______________________ Score ______________________

Activity 4-8
Transportation Technology

Type of hull: ______________________

On the grid below, draw a sketch of the hull you will build for your boat. Dimension the major distances.

Hull Test Data

Record the data from the tests of the three hull shapes. Rank them in terms of their efficiency.

Hull Design #1 Type of design: ____________________

Sketch shape below.

Travel time:

Test #1 __________

Test #2 __________

Test #3 __________

Average __________

Rank

Hull Design #2 Type of design: ____________________

Sketch shape below.

Travel time:

Test #1 __________

Test #2 __________

Test #3 __________

Average __________

Rank

Hull Design #3 Type of design: ____________________

Sketch shape below.

Travel time:

Test #1 __________

Test #2 __________

Test #3 __________

Average __________

Rank

Name ______________________ Date ______________________

Period ______________________ Score ______________________

Activity 4-9
Hovercraft Design

The Challenge

Design and test a model of hovercraft, using the drawing below or your own ideas to develop your plans.

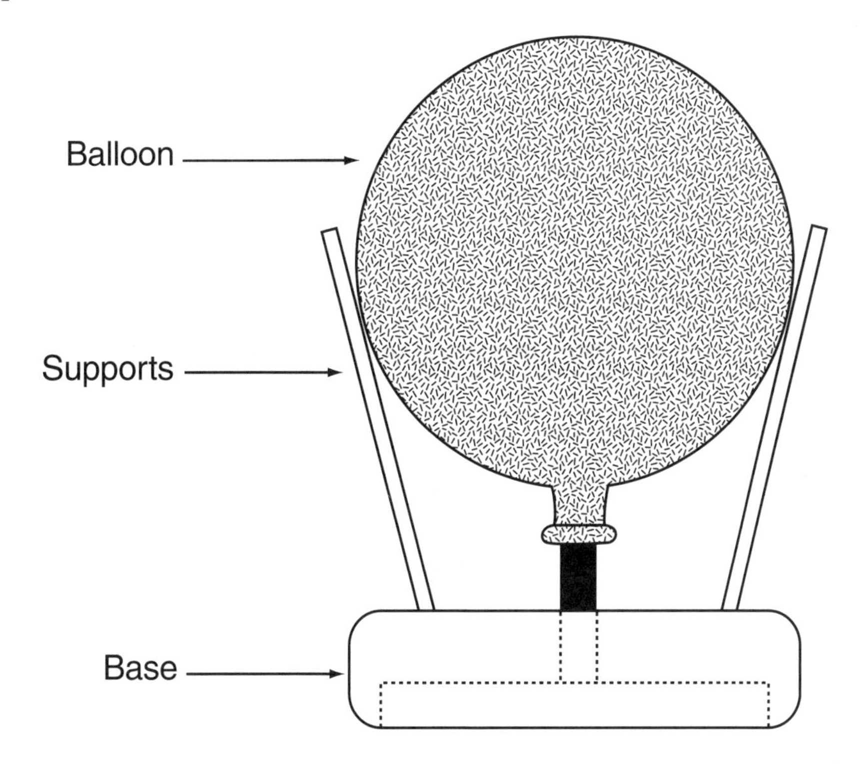

Background

In some situations, a vehicle must travel over a land surface that will not support its weight. This condition gave rise to air-cushion vehicles (ACVs). These vehicles are suspended on cushions of air. The air is forced downward, under the vehicle. The force of the air causes the vehicle to rise slightly, allowing the air to escape from under the vehicle. A second blast of air in the opposite direction of desired travel will move the vehicle along its pathway.

Design Brief

Using only a piece of 3/4" × 4" × 4" extruded polystyrene insulation board, one 1 1/4" length of 3/8" OD air hose, a balloon, and four soda straws, construct a vehicle that will ride on a cushion of air.

Procedure

Design and test your ACV using the following procedure:

1. Prepare a sketch of your design.
2. Dimension and label all the parts on the sketch.
3. Build a prototype of the vehicle.
4. Test the model.
5. Evaluate the model's performance.
6. Revise the sketch to improve the model.
7. If time permits, construct and test a revised model.

Additional Challenges

1. Develop a system to control the airflow under the vehicle.
2. Develop an additional power source to propel the vehicle forward.
3. Design a system that can be used to steer the vehicle along a curved path.

Name ______________________ Date ______________

Period ______________________ Score ______________

Chapter 19
Study Questions

Agricultural and Related Technology

Textbook pages: 434–465

1. Modern agriculture uses both science and technology. True or false. ______________
2. Modern farming machines and better ways to store food are part of agricultural _______. ______________
3. Name six common types of agricultural crops. ______________

____

____

____

____

____

Match the statement on the left with the farm equipment it describes on the right.

4. Breaks and pulverizes the soil. ______________
5. Cultivates soil to remove weeds. ______________
6. Has an engine, transmission, traction device, and hitch. ______________
7. Sows seeds. ______________
8. Cuts and separates grains from stalks. ______________
9. Protects harvested crops from the weather. ______________
10. Provides crops with water during dry spells. ______________

A. Power or pulling.
B. Tilling.
C. Planting.
D. Pest control.
E. Irrigation.
F. Harvesting.
G. Storage.

11. Growing crops in nutrient solutions without soil is called __________. ______________________________

12. Aquaculture involves growing and harvesting __________ in controlled conditions. ______________________________

13. __________ is using biological agents in processes to produce goods and services. ______________________________

Name ______________________ Date ______________________

Period ______________________ Score ______________________

Chapter 20
Study Questions

Construction Technology

Textbook pages: 466–491

1. Name five common types of buildings. ______________________
____
____
____
____

2. Bridges, roads, and dams are examples of __________ structures. ______________________

3. Architectural __________ show the shape and size of a proposed structure. ______________________

4. Public construction projects are paid for with money obtained through __________. ______________________

5. The edges of a piece of property are called its __________. ______________________

6. The distance a building must be in from a property boundary is set by __________ restrictions or codes. ______________________

7. A(n) __________ is a substructure upon which a building is built. ______________________

8. What are the three main parts of a building's frame? ______________________
____
____

9. The vertical lumber parts of a wall are called its __________. ______________________

10. The systems providing fresh water and removing wastewater from a home are __________ systems. ______________________

11. __________ keeps outside heat from entering a home. ____________________

12. __________ includes planting trees, shrubs, and lawns. ____________________

Name ______________________ Date ______________________

Period ______________________ Score ______________________

Chapter 21
Study Questions

Energy Conversion Technology

Textbook pages: 492–513

1. The ability to do work is called _________. ______________________
2. Name the five major types of energy converters. ______________________

____

____

____

____

3. An electric generator is a(n) _________ energy converter. ______________________
4. Electricity moving first in one direction and then in the opposite direction is called _________ current. ______________________
5. An internal combustion engine is an example of a(n) _________ energy converter. ______________________
6. Jet engines and gas turbines are types of _________ propulsion devices. ______________________
7. Name the two types of rocket engines. ______________________

____

8. _________ are the most common devices changing chemical energy into electrical energy. ______________________
9. A(n) _________ produces electricity from a fuel and oxygen. ______________________
10. Name the two types of solar converters. ______________________
11. A(n) _________ uses a sail to convert wind energy into mechanical energy. ______________________
12. Shafts, pulleys, and gears are examples of _________ transmission of energy. ______________________

Name ______________________ Date ______________
Period ______________________ Score ______________

Chapter 22
Study Questions

Information and Communication Technology

Textbook pages: 514–555

Match the statement on the left with the term it describes on the right. (Answers may be used more than once.)

1. Sorted facts and figures. ______
2. Organized facts and opinions. ______
3. Individual facts, statistics, and ideas. ______

A. Data.
B. Information.

4. Name the two major ways people communicate using technology. ______ ______
5. Using drawings, printed words, and pictures is called _____ communication. ______
6. Radio and television are examples of _____ communication. ______
7. The channel carrying a message is called a(n) _____. ______
8. The word _____ means communicating over a distance. ______
9. The language used to format documents on the World Wide Web is called _____. ______

Match the statement on the left with the printing process it describes on the right. *(continued on back)*

10. Prints from recessed images (engravings). ______
11. Forces ink through fabric. ______
12. Photocopying. ______
13. Prints from raised surfaces on type. ______

A. Relief.
B. Offset lithography.
C. Intaglio.
D. Screen printing.
E. Electrostatic.

14. Transfers an inked image on a plate to a blanket and then to the paper. ______________

15 Name the parts of the simple communication system shown below.

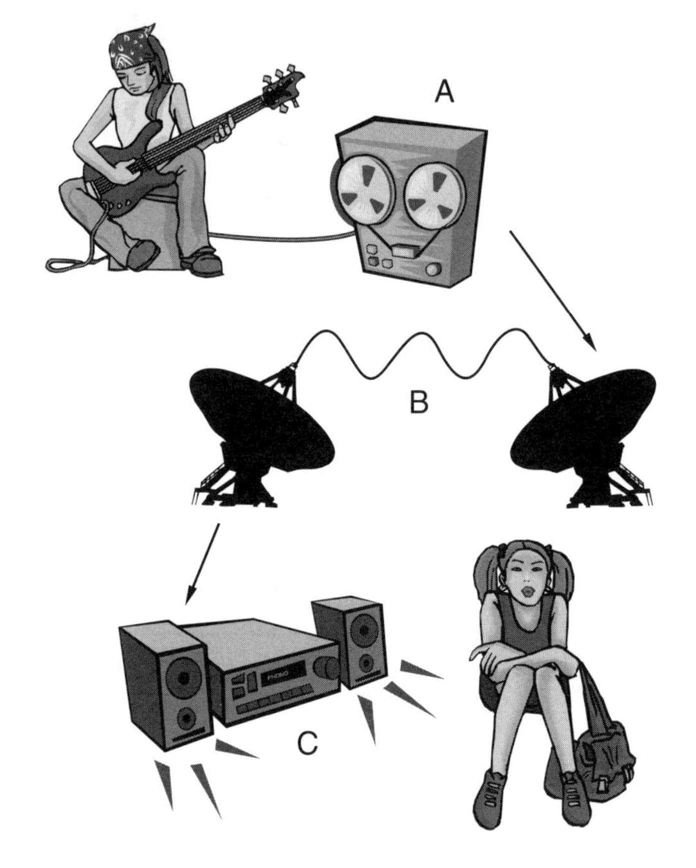

A ______________

B ______________

C ______________

Match the statement or term on the left with the part of a computer it describes on the right. (Answers may be used more than once.)

16. "Brains" of the computer. ______________
17. Keyboard. ______________
18. CD-ROM. ______________
19. Hard drive. ______________
20. Scanner. ______________
21. RAM. ______________
22. Printer. ______________
23. Processes information. ______________

A. Input device.
B. Central processing unit.
C. Memory.
D. Output device.

Name ______________________ Date ______________

Period ______________________ Score ______________

Chapter 23
Study Questions

Manufacturing Technology

Textbook pages: 556–581

Match the statement on the left with the term it describes on the right.

1. Manufacturing changes _________ materials into _________ products. ______________ ______________
2. Changing trees into lumber is an example of _________ processing. ______________
3. Changing lumber into a bookcase is an example of _________. ______________
4. Name the three major types of primary processes. ______________ ______________ ______________

Match the statement on the left with the process it describes on the right. Answers may be used more than once. *(continued on back)*

5. Applying a coating to protect the material. ______________
6. Pouring a molten material into a mold. ______________
7. Bolting parts together. ______________
8. Applying pressure to force a material over a die. ______________
9. Giving shape by removing excess material. ______________
10. Includes forging and thermoforming. ______________
11. Designed to change the property of the material. ______________
12. Includes organic and inorganic coatings. ______________
13. Includes shearing and machining. ______________

A. Casting and molding.
B. Forming.
C. Separating.
D. Conditioning.
E. Assembling.
F. Finishing.

14. Uses permanent and expendable (one-shot) molds. __________

15. Name four major types of manufacturing systems. __________ __________ __________ __________

16. __________ charts list the operation, transportation, inspection, and storage activities needed to build a product. __________

17. Special devices holding and positioning parts to make manufacturing more efficient are called __________. __________

18. Checking parts for quality is called __________. __________

Match the statement on the left with the management function it describes on the right. (Answers may be used more than once.)

A. Planning.
B. Organizing.
C. Actuating.
D. Controlling.

19. Comparing results with the plans. __________

20. Assigning jobs. __________

21. Setting goals. __________

22. Dividing tasks into jobs. __________

23. Supervising workers. __________

24. __________ tells people about a product and its benefits. __________

Name ______________________ Date ______________________

Period ______________________ Score ______________________

Chapter 24
Study Questions

Medical Technology

Textbook pages: 582–603

1. Name the three major goals of health and medical programs. ______________________

____

____

2. __________ is a state of well being. ______________________

3. What are the four important factors in a wellness program? ______________________

____

____

____

Match the statement on the left with the term it describes on the right. (Answers may be used more than once.)

4. Uses a large number of different muscles.	A. Anaerobic exercise.	______________
	B. Aerobic exercise.	
5. Increases strength and muscle mass.		______________
6. Weight lifting and sprinting.		______________
7. Reduces chance of heart disease.		______________
8. Increases endurance.		______________

9. _____ uses modified or killed microorganisms to develop natural resistance to disease. ______________________

10. Name the two major steps in dealing with an illness or a medical condition. ______________________

____

Match the statement on the left with the diagnostic device it describes on the right. (Not all answers will be used.)

A. X ray.

B. CT scanner.

C. MRI.

D. Ultrasound.

11. Uses high frequency sound waves and their echoes to develop an image of the body. ______________

12. Uses short electromagnetic waves to develop a two-dimensional image on film. ______________

13. Uses magnetic waves to create an image. ______________

14. Taking tissue samples for laboratory testing is called __________. ______________

15. __________ uses high-energy radiation to treat cancer. ______________

16. Damaged or diseased organs can be removed through __________. ______________

17. __________ are special chemicals used to treat illnesses. ______________

18. __________ are special chemicals that help develop immunity against a disease. ______________

Name ______________________ Date ______________

Period ______________________ Score ______________

Chapter 25
Study Questions

Transportation Technology

Textbook pages: 604–627

How many degrees of freedom do each of the following transportation devices have?

1. Automobile. ______
2. Railroad engine. ______
3. Pipeline. ______
4. Airplane. ______
5. Cruise liner. ______
6. Submarine. ______
7. A city bus is an example of a(n) _________ transportation system. ______________
8. A family car is an example of a(n) _________ transportation system. ______________
9. Name the six transportation processes or actions. ______________

 ____

 ____

 ____

 ____

 ____

10. Land transportation systems that have one degree of freedom are called _________ systems, while those with two degrees of freedom are called _________ systems. ______________

 ____

11. _________ trains carry large quantities of one material. ______________
12. Pathways ships follow across the oceans are called _________. ______________

Match the statement on the left with the vehicle system it describes on the right. (Answers may be used more than once.)

13. Enables the vehicle to change speed. _____________________________

14. Maintains the vehicle on the pathway. _____________________________

15. Provides the framework for the vehicle. _____________________________

16. Enables the vehicle to change direction. _____________________________

17. Enables the vehicle to move. _____________________________

18. Receives information to operate the vehicle. _____________________________

A. Structure.
B. Propulsion.
C. Suspension.
D. Guidance.
E. Control.

19. A __________ is a vehicle riding on a cushion of air. _____________________________

20. Terminals, bridges, and roadways are examples of __________ for transportation activities. _____________________________

Section 5

Technology and Society

Name ______________________ Date ______________________

Period ______________________ Score ______________________

Activity 5-1
Future Telling

Complete the form below by creating a future scenario. Begin by selecting a future event. The event should be something you think, or hope, will happen in the future. It can be a future problem or possible situation. Next, list possible solutions. Lastly, list the outcomes of the event. Write a short essay on how the people of the future will react to the event. Be as creative as you can.

Future Event

(Example: Every student in the world wants to take technology education classes every year.)

____

____

Possible Solutions

(Example: More technology teachers are hired or technology is taught in other classes, as well as in technology education classes.)

____

____

____

____

Outcomes and Reactions

____

____

____

____

____

Name ______________________________ Date ____________________

Period ______________________________ Score ____________________

Activity 5-2
Products of the Future

Group #: _____

Members: ____

____

Circle your assigned application of technology.

Agricultural and bio-related
Construction
Energy conversion
Information and communication
Manufacturing
Medical
Transportation

Write several current examples of products and systems in your application.

____

____

____

____

Brainstorm possible technologies that will exist in the future (in your application). Make notes of your brainstorming session below.

____

____

____

____

Circle your best idea from the notes above.

Drawing

In the space below, draw your future technology. Label any parts needing to be explained. Color and decorate as needed.

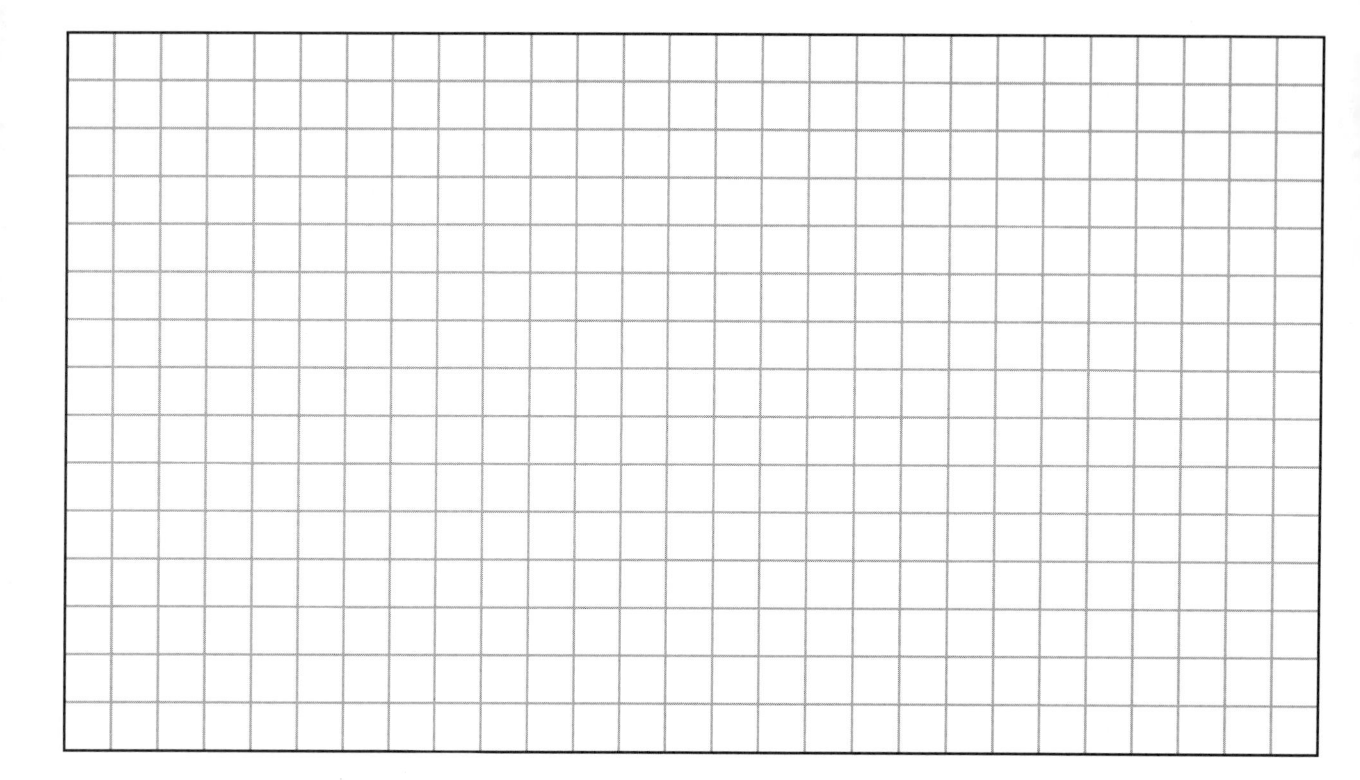

Explanation

Explain the features of your future technology. Explain what it does, how it works, who will use it, and in what year it will be developed.

Name ______________________ Date ______________________

Period ______________________ Score ______________________

Chapter 26
Study Questions

Technological Impacts

Textbook pages: 654–677

1. Technological advancements cause _________. ______________________

Match the technology on the left with the type of change it caused on the right.

2. Plastic. A. Construction. ______________________
3. Internet. B. Communication. ______________________
4. Skyscraper. C. Materials. ______________________
5. Cordless drill. D. Tools. ______________________
6. List three major impacts technology has on society. ______________________

____

____

Name ______________________ Date ______________

Period ______________________ Score ______________

Chapter 27
Study Questions

Technology and the Future

Textbook pages: 678–691

1. A(n) __________ is an expert who studies the future. ______________
2. A long-range projection may look ahead as far as __________ years. ______________

Match the definition on the left with the term on the right.

3. Studying the direction in which events are going. ______________
4. Asking experts to respond to a number of questions about the future. ______________
5. Creating an outline of future events. ______________

A. Surveying.
B. Trending.
C. Developing a scenario.

6. The use of color X rays will be an advance in __________ technology. ______________
7. The future use of robots will be important in __________ technology. ______________

Communication Guidelines

The following sections provide guidelines in a variety of communication media. Following these guidelines will help to ensure that your communication messages meet the receiver's expectations.

Word Processing

Probably the most common form of communication you will encounter in school and in the field of technology is written communication. Most written communication is produced using word processing. Word processors are computer software programs that allow you to create documents that are neat and easy to read. It is up to you to make sure that the content is organized and accurate.

When typing a document, there are several guidelines to follow. First, always use correct grammar and punctuation. If you are unsure about what is correct, use a dictionary or other resource to find out. Next, avoid using extra returns to skip lines between paragraphs. Similarly, use only one space after each sentence. To modify the visual effects of your document even more, you can change the font type or size, justify the text, add tabs, or use a different text style.

Word processors help you create electronic files, and these files should be carefully saved for later use. Normally, they can be saved on a floppy diskette or on your computer's hard drive. It is wise to save your work often and to even save a backup copy so you are sure not to lose it.

Communicating with written documents created on a word processor is most effective if your message is organized, accurate, and clear. For this reason, it is important to proofread your work before finalizing it. Often, several drafts need to be written before the document is acceptable for printing. Once the document is finished, proofread, edited, and saved, you can usually print it out on a printer connected to your computer.

Video Communication

Like any other form of communication, video requires planning and organization. When outlining your message, remember that videos are largely visual. The pictures you film are often more important than the words you write. To help organize your visual images, it is useful to create a storyboard, which is a series of pictures that tell a story, much like a comic strip. Once your pictures are created, you can try rearranging them for different effects and decide on a final sequence to use in your video. In addition to visual images, your video will probably also require a script. The spoken words follow the sequence of pictures and help tell your message.

After the initial planning is complete, it is time to begin planning for the production of your video. Scenes are not always recorded in the sequence in which they will appear in the final product. The order in which you film the scenes may depend on the availability of set locations, groups of actors, and special props. To help ensure that your final product is interesting to watch, you should plan to record your subject from various angles and using a wide range of shots. Also try to shoot cutaway shots, which are filler shots that can be used for transitions.

Video communication does require specialized equipment. There are many types of video cameras, and you will want to use one that best suits your needs. Extra batteries and hand-held microphones may also be needed.

Once you have recorded your scenes, there are numerous effects you can use to enhance your video message. Video production software is useful for applying some of these effects to your recording. Both visual effects and sound effects can be added to your video, and voice-over narration can also be added during post-production. Your cutaway shots can also be inserted to create a smooth, flowing final product.

PowerPoint Communication

Oral presentations can be greatly enhanced by incorporating a PowerPoint presentation as well. Content and design are both important when dealing with PowerPoint, so it is essential that you research your topic well and plan out a general design for your PowerPoint slides. Your slide design should be kept simple, and the font size should be large enough to read from the back of the room where you will be giving your presentation. Each slide should include a title and a single piece of information or a short list of bulleted points. The information you present in your oral presentation should only be outlined on the PowerPoint slides, not written word for word.

Web-Based Communication

Web-based communication is becoming increasingly popular and important in technological industries. Developing a website takes a good deal of planning and time. Sketch a plan for each web page, and make sure you include all the necessary links between pages.

The most important characteristics of a good website are ease of navigation and functionality. It is important that your site has a nice visual appearance, but it is more important that the type is easy to read and the links are easy to find. Use common fonts and limit the types and sizes of fonts as much as possible. You can add images where they may be appropriate, but using too many images will cause the page to load slowly and may detract from your intended message. To create a successful website, try to keep things simple and clear.

Desktop Publishing Communication

Communicating with desktop publishing software is a strictly visual activity, so the principles of design are especially important. When designing a message for printed communication, the various elements on the message should be balanced and proportionate. Contrast, unity, and rhythm can be used to achieve the desired effect.

Font types and sizes should be kept to a minimum to avoid confusing your audience. Choose your illustrations carefully and do not forget the effects that the use of white space can have on your final product. Try arranging the elements of your design in several ways before deciding which way would be most effective for your purpose. Make sure to draw particular attention to any titles or other essential information. The visual design of your message will determine how the reader's eyes will move when viewing your finished product.